# An Illustrated Data Guide To

# Modern Aircraft Carriers

Compiled by
Christopher Chant

**TIGER BOOKS INTERNATIONAL
LONDON**

This edition published in 1997 by
Tiger Books International PLC
Twickenham

Published in Canada in 1997 by
Vanwell Publishing Limited
St. Catharines, Ontario

© Graham Beehag Books
Christchurch
Dorset

All rights reserved. No part of this book may be reproduced or transmitted in any form or by any means electronic or mechanical including photocopying, recording or any information storage system without permission in writing from the Publisher.

Printed and bound in Hong Kong

ISBN 1-85501-863-2

# Contents

Charles de Gaulle class ...................... 8
Clemenceau class ............................. 11
Colossus class .................................. 16
Enterprise class ................................ 21
Forrestal class .................................. 25
Garibaldi class ................................. 30
Hermes class ................................... 34
Invincible class ................................ 37
John F. Kennedy class ...................... 44
Kiev class ........................................ 48
Kiev (Modified) class ........................ 55
Kitty Hawk class ............................... 57
Kuznetsov class ............................... 60
Majestic class .................................. 62
Midway class ................................... 65
Moskva class ................................... 67
Nimitz class ..................................... 70
Principe de Asturias class ................ 76

# MODERN AIRCRAFT CARRIERS

## 'Charles de Gaulle' class

**Country of origin:** France

**Type:** Nuclear-powered multi-role aircraft carrier

**Displacement:** 34,500 tons standard and 36,600 tons full load

**Dimensions:** Length 857ft 11in (261.5m); beam 103ft 4in (31.5m) and width 211ft 4in (64.4m); draught 27ft 11in (8.5m); flightdeck length 857ft 11in (261.5m) and width 211ft 4in (64.4m)

**Gun armament:** Eight 20mm 20F2 AA in four twin mountings

**Missile armament:** Four SAAM octuple vertical-launch systems for SAN 90 Aster 15 surface-to-air missiles, and two SADRAL sextuple launchers for Mistral surface-to-air missiles

**Torpedo armament:** None

**Anti-submarine armament:** Aircraft and helicopters (see below)

# CHARLES DE GAULLE CLASS

**Aircraft:** 40 fixed-wing and an unknown number of rotary-wing machines

**Electronics:** One DRBJ 11D/E 3D surveillance radar, one DRBV 26D air-search radar, one DRBV 15C air/surface-search radar, two Arabel 3D air-search and Aster fire-control radars, two Racal 1229 navigation radars, one NRBA-series landing radar, one Syracuse satellite navigation system, one torpedo-warning sonar, one SENIT 6 action information system, one ESM system with one ARBR 17 warning and two ARBB 33 jamming elements, one DIBV 10 Vampir IR detector, four Sagaie chaff/flare launchers, one satellite communication system, and NRBP 6A TACAN

**Propulsion:** Two Type K15 pressurised water-cooled reactors supplying steam to two sets of geared turbines delivering 83,020hp (61,900kW) to two shafts

**Performance:** Maximum speed 27kt

**Complement:** 1,150 plus an air group of 550, with accommodation for 1,950 possible

*Seen under construction at Brest Naval Dockyard, the Charles de Gaulle is the first of two required as replacements for the elderly 'Clemenceau' class carriers, but construction has been slowed for financial reasons.*

# MODERN AIRCRAFT CARRIERS

### France

| Name | No. | Builder | Commissioned |
|---|---|---|---|
| Charles de Gaulle | R91 | Brest ND | 1998 |
| Richelieu | R92 | Brest ND | 2004 |

**Note**

Designed to replace the two 'Clemenceau' carriers at about the turn of the century, these carriers will each have a flightdeck angled at 8.5°, and will be fitted with two 246ft (75m) steam catapults and two aircraft lifts each measuring 62ft 4in (19.0m) by 41ft 0in (12.5m). The ships use a reactor design derived from that of the French SSBN force, and the comparatively low power thus imposed makes these ships the slowest aircraft carriers in the world apart from the Spanish *Principe de Asturias*.

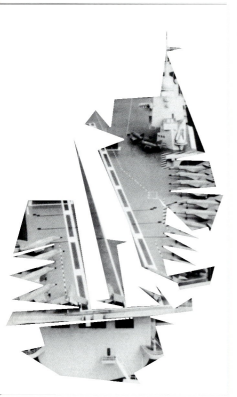

This is a tank test model of the Charles de Gaulle, *but reveals the nature of the flightdeck arrangement with parking area to port and starboard of the angled section of the deck, which is fitted with a single steam catapult, a straight take-off section forward with its own steam catapult, and the large starboard-side island forward of the two deck-edge lifts. The ship's defensive armament is scattered along the port and starboard sides of the flightdeck sections.*

## CLEMENCEAU CLASS

# 'Clemenceau' class

**Country of origin:** France

**Type:** Conventionally powered multi-role aircraft carrier

**Displacement:** 27,310 tons standard and 32,780 tons full load

**Dimensions:** Length 869ft 5in (265.0m); beam 104ft 0in (31.7m) and width 168ft 0in (51.2m); draught 28ft 3in (8.6m); flightdeck length 843ft 2in (257.0m) and width 168ft 0in (51.2m)

**Gun armament:** Four 3.94in (100mm) Creusot Loire L/55 DP in Modèle 1953 single mountings

**Missile armament:** Two Naval Crotale EDIR octuple launchers for 36 Matra R.440 surface-to-air missiles

**Torpedo armament:** None

**Anti-submarine armament:** Aircraft and helicopters (see below)

**Aircraft:** Typically 36 fixed-wing (16 Dassault-Breguet Super Etendard, 3 Dassault-Breguet Etendard IVP, 10 Vought F-8 Crusader and 7 Dassault-Breguet Alizé) machines and 2 rotary-wing (Aérospatiale SA 365F Dauphin 2) machines

**Armour:** Flightdeck, hull sides and bulkheads (magazine and engine room areas), and island

**Electronics:** One DRBV 23B surveillance radar, two DRBI 10 air/surface-search radars, one DRBV 15 air-warning radar, two DRBC 32 surface-to-air fire-control radars, one Decca 1226 navigation radar, one NRBA 51 landing radar, one SQS-505 active search hull sonar, one SENIT 2 action information system, one ESM system with ARBR 16 warning and ARBX 10 jamming elements, two Sagaie chaff/flare launchers, and SRN-6 TACAN

**Propulsion:** Six boilers supplying steam to two sets of Parsons geared turbines delivering 126,000hp (93,960kW) to two shafts

# MODERN AIRCRAFT CARRIERS

**Performance:** Maximum speed 32kt; range 8,635 miles (13,900km) at 18kt or 4,040 miles (6,500km) at 32kt

**Complement:** 64+1,274

| France | | | |
|---|---|---|---|
| Name | No. | Builder | Commissioned |
| Clemenceau | R98 | Brest ND | Nov 1961 |
| Foch | R99 | Ch. de l'Atlantique | Jul 1963 |

# CLEMENCEAU CLASS

*The* Clemenceau, *name ship of a two-ship class, is an obsolescent aircraft carrier that reflects the 'state of the art' at the time of her construction in the late 1950s, although she has been updated since that time. Her comparatively small overall size and modestly dimensioned flightdeck precludes the carriage of a large air wing or the operation of modern warplanes.*

### Note

The *Clemenceau* was the first aircraft carrier to be designed as such in France and actually completed. Built in the late 1950s and commissioned in November 1961, she incorporated all the advances made in carrier design during the early 1950s, namely a fully angled flightdeck, mirror landing sight and a fully comprehensive set of air-search, tracking and air-control radars. The part of the flightdeck angled at 8° to port off the ship's centreline measures 543ft 0in (165.5m) in length and 96ft 9in (29.5m) in width, and the flightdeck is served by two aircraft lifts, each rated at 44,896lb (20,365kg) and measuring 52ft 6in (16.0m) by 36ft 1in (11.0m), as one unit abaft the island on the deck edge and the other offset to starboard and just forward of the island. The flightdeck is fitted with two steam catapults (one each on the angled and bow sections). The hangar has a usable volume of 498ft 8in (152.0m) in length by 78ft 9in (24.0m) in width by 23ft 0in (7.0m) in height. The fuel capacity of the *Clemenceau* is 42,377cu ft (1,200cu m) of JP5 aircraft fuel and 14,126cu ft (400cu m) of AVGAS, while the *Foch* carries 63,566 and 1,173cu ft (1,800 and 109cu m) respectively of these commodities.

Between September 1971 and November 1918 the *Clemenceau* underwent a major refit, the *Foch* following her between July 1980 and August 1981. In the course of these refits both ships were converted to operate the Dassault-Breguet (now Dassault) Super Etendard strike fighter, for which the ships embark 15kT AN-52 tactical nuclear bombs. The ships also received SENIT 2 automated tactical information-processing systems as part of their command, control and communication suites. The two carriers' air groups now each comprise 16 Super Etendard strike fighters, three Dassault (originally Dassault-Breguet) Etendard IVP photographic reconnaissance fighters, 10 Vought F-8E Crusader interceptors and seven Dassault

(originally Breguet) Alizé ASW aircraft, plus two Aérospatiale Super Frelon ASW and two Aérospatiale Alouette III utility helicopters.

If required by the operational situation, the carriers can also act as helicopter carriers with an air group of 30–40 helicopters depending upon the types embarked. During the Lebanon crisis of 1983, France used one of the carriers in support of her peace-keeping force, Super Etendards being used to attack several Moslem gun positions that had engaged French troops. It had been planned that the *Clemenceau* would be paid off in 1992 and the *Foch* in 1998 as they were replaced by the two nuclear-powered units of the 'Charles de Gaulle' class, but the programme to build the new class has been delayed by the end of the Cold War and by retrenchment in military spending, so the two 'Clemenceau' class ships will survive in French service into the next century and there is now a possibility that at least one may be bought by China.

*The* Clemenceau *(below left) and the* Foch *(below) were named after France's political and military leaders of World War I, and both ships are now obsolescent technically as well as being increasingly difficult to maintain. Each ship had a moderately angled flightdeck section, two Mitchell-Brown steam catapults each able to accelerate a 44,092lb (20,000kg) warplane to 110kt (127mph; 204km/h), and two aircraft lifts located as one unit in the flightdeck and the other on the starboard edge of the flightdeck. The typical air wing comprises 36 fixed-wing and two rotary-wing aircraft.*

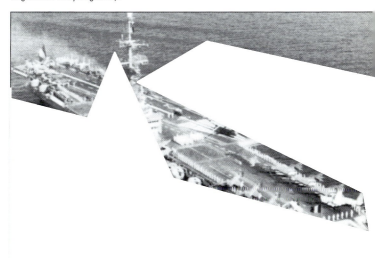

# MODERN AIRCRAFT CARRIERS

## 'Colossus' class

**Country of origin:** UK, now Brazil

**Type:** Conventionally powered light anti-submarine aircraft carrier

**Displacement:** 15,890 tons standard and 19,890 tons full load

**Dimensions:** Length 695ft 0in (211.8m); beam 80ft 0in (24.4m); draught 24ft 6in (7.5m); flightdeck length 690ft 0in (210.3m) and width 119ft 7in (36.4m)

**Gun armament:** Ten 40mm Bofors L/60 AA in two Mk 2 quadruple mountings and one Mk 1 twin mounting

**Missile armament:** None

**Torpedo armament:** None

**Anti-submarine armament:** Aircraft and helicopters (see below)

**Aircraft:** Typically 6 fixed-wing (Grumman S-2E Tracker) machines and 11 rotary-wing (6 Sikorsky SH-3D Sea King, 3 Aérospatiale AS 332F Super Puma and 2 Aérospatiale AS 355F Ecureuil 2) machines

*The Minas Gerais is used for the anti-submarine role, possesses only a short angled section at the rear of the flightdeck, and is fitted with one MacTaggart-Scott steam catapult. The ship's role is reflected in the typical air-wing composition of six Grumman S-2 Tracker fixed-wing anti-submarine aircraft and 11 helicopters, including up to six examples of the Sikorsky SH-3 Sea King and three Eurocopter France (Aérospatiale) Super Puma helicopters.*

## COLOSSUS CLASS

**Electronics:** One SPS-40B air-search radar, one SPS-4 surface-search radar, one SPS-8A combat-control radar, one Raytheon MP 1402 navigation radar, two SPG-34 40mm fire-control radars, two Mk 63 gun fire-control systems, one Mk 51 gun fire-control system, one SLR-2 ESM system with warning element, and TACAN

**Propulsion:** Four Admiralty boilers supplying steam to two sets of Parsons geared turbines delivering 40,000hp (29,830kW) to two shafts

**Performance:** Maximum speed 24kt; range 13,800 miles (22,210km) at 14kt or 7,140 miles (11,490km) at 23kt

**Complement:** 1,000 plus an air group of about 300

| Argentina | | | |
|---|---|---|---|
| *Name* | *No.* | *Builder* | *Commissioned* |
| *Veinticinco de Mayo* | V2 | Cammell Laird | Jan 1945 |

**Note**
The *Veinticinco de Mayo* (25 May) was originally a 'Colossus' class aircraft carrier purchased from the United Kingdom by the Dutch and commissioned into the Royal Netherlands Navy in May 1948. In April 1968 the ship suffered a serious fire in her boiler rooms and was subsequently judged to be

## MODERN AIRCRAFT CARRIERS

beyond economic repair. In October, Argentina bought the vessel, which was refitted and commissioned into the Argentine navy while in the Netherlands, and then sailed for Argentina in September 1969. The vessel is fitted with a modified Ferranti CAAIS data-processing system and Plessey Super CAAIS console displays: this combination allows the ship to control her aircraft and to communicate via data-links with the two British-built 'Type 42' class destroyers of the Argentine navy and their ASAWS 4 action information systems. The *Veinticinco de Mayo*'s modified superstructure differs considerably from those of ex-British carriers in service with other navies.

In 1980-81 the ship underwent a further refit to increase the strength of the flightdeck and to add deck space for two additional aircraft, in readiness for the Dassault Super Etendards that Argentina was acquiring. None of these strike aircraft had qualified to land on the carrier by the time of the Falklands war in 1982, and the carrier's air group consisted of eight McDonnell Douglas A-4Q Skyhawks, six Grumman S-2E Trackers and four Sikorsky SH-3D Sea Kings. The *Veinticinco de Mayo* played a major part in the initial landings on the Falklands, and was ready to launch a strike against the British task force at the beginning of May 1982, but poor flying conditions forced the

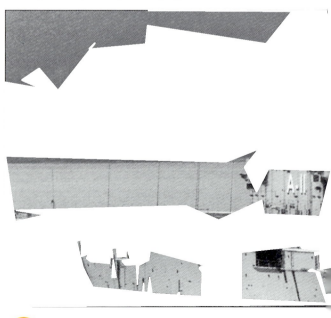

cancellation of the mission. The subsequent sinking of the Argentine 'Brooklyn' class cruiser *Belgrano* by a British nuclear-powered submarine then forced the Argentine naval command to pull the carrier back to the relative safety of Argentina's coastal waters, where the ship played no further part in the war and landed her air group for land-based operations.

After the Argentine loss of the Falklands the remaining Super Etendards were delivered. These were rapidly deck-qualified and the new complement of the air group is described as 18 fixed-wing aircraft (12 Dassault-Breguet Super Etendards and six Grumman S-2E Trackers, although the McDonnell Douglas A-4P/Q Skyhawk can also be embarked) and five rotary-wing aircraft (four Sikorsky SH-3D Sea King and one Aérospatiale SA 319B Alouette III). The electronic fit comprises one LW-08 surveillance radar, one DA-08 air/surface-search radar, one VI/SGR-109 height-finding radar, one DA-02 surface-search radar, one ZW-01 navigation radar, one SPN 720 landing radar, URN-20 TACAN, and one CAAIS action information system. The armament comprises nine 40mm Bofors L/70 AA in single mountings. In the late 1980s the ship was taken in hand for revision with diesel engines in place of the elderly Parsons steam turbines: this change will ease maintenance and

*This ship, now known as the Minas Gerais, was completed as the Vengeance for the Royal Navy in January 1945, was lent to the Royal Australian Navy between 1953 and 1955, and was then sold to the Brazilian navy in 1956. The subsequent modernisation effort, completed in the Netherlands, saw the addition of the angled flightdeck section, mirror-sight deck landing system, current steam catapult and aircraft lifts, increased steam capacity, and a remodelling of the island.*

# Modern Aircraft Carriers

improve reliability, and will enable a higher speed so that the ship is more compatible with its Super Etendard strike fighters, which have often been forced to operate from shore bases.

### Brazil

| Name | No. | Builder | Commissioned |
|------|-----|---------|--------------|
| Minas Gerais | A11 | Swan Hunter | Jan 1945 |

**Note**

A half-sister of the Argentine *Veinticinco de Mayo*, this ship was completed as the *Vengeance* for service in the Royal Navy from 1945. Three years later the ship was fitted out for an experimental cruise to the Arctic and was then lent to the Royal Australian Navy in 1953. She was returned to the Royal Navy in 1955 and was purchased by Brazil in December 1956 as the *Minas Gerais*. She was transferred to the Netherlands, where she was comprehensively refitted between 1957 and 1960 to a standard that included new weapons, a steam catapult of 29,465lb (13,365kg) capacity, an 8.5° angled flightdeck, a mirror-sight deck landing system, a new island superstructure, new American radars, and two centreline aircraft lifts each measuring 45ft 0in (13.7m) by 34ft 0in (10.4m). The hangar is 445ft 0in (135.6m) long by 52ft 0in (15.8m) wide by 17ft 6in (5.3m) high.

In 1976-81 the carrier underwent another refit to allow her to operate through to the 1990s. A data-link system was installed so that the carrier can co-operate with the 'Niteroi' class of frigates in service with the Brazilian navy, and the obsolete American SPS-12 radar was replaced by a modern SPS-40B 2D air-search system. Throughout her service with the Brazilian navy, the *Minas Gerais* has been tasked with the anti-submarine warfare role with an air group (since the late 1970s) of eight Grumman S-2E Trackers of the Brazilian air force (by law, the Brazilian navy is not being permitted to operate fixed-wing aircraft), plus four navy Sikorsky SH-3D Sea King ASW helicopters as well as two Aérospatiale SA530 Ecureuil and two Bell Model 206 JetRanger utility helicopters. It is thought that the Brazilian navy would like to replace the *Minas Gerais* with two carriers capable of operating a mixed air group of STOVL fighter/strike aircraft and ASW helicopters.

# 'Enterprise' class

**Country of origin:** USA

**Type:** Nuclear-powered multi-role aircraft carrier

**Displacement:** 75,700 tons standard and 90,970 tons full load

**Dimensions:** Length 1,088ft 0in (331.6m); beam 133ft 0in (40.5m); draught 39ft 0in (11.9m); flightdeck length 1,088ft 0in (331.6m) and width 252ft 0in (76.8m)

**Gun armament:** Three 20mm Phalanx Mk 15 CIWS mountings, and three 20mm AA in Mk 68 single mountings

# MODERN AIRCRAFT CARRIERS

**Missile armament:** Three Mk 29 octuple launchers for RIM-7 NATO Sea Sparrow surface-to-air missiles

**Torpedo armament:** None

**Anti-submarine armament:** Aircraft and helicopters (see below)

**Aircraft:** Typically 90 in a multi-role carrier air wing with 20 Grumman F-14 Tomcat, 20 McDonnell Douglas F/A-18 Hornet, 5 Grumman EA-6B Prowler, 20 Grumman A-6E Intruder, 4 Grumman KA-6D Intruder, 5 Grumman E-2C Hawkeye and 10 Lockheed S-3A Viking fixed-wing aircraft, and 6 Sikorsky SH-3G/H Sea King helicopters

**Electronics:** One SPS-48E 3D radar, one SPS-49(V)5 air-search radar, one SPS-65 threat-warning radar, one SPS-58 low-level threat-warning radar, one SPS-67 surface-search radar, one SPN-64(V)9 navigation radar, five SPN-series aircraft landing radars, one Mk 23 and six Mk 57 radars used in conjunction with three Mk 91 SAM fire-control systems, one Naval Tactical Data System (NTDS), Links 4A, 11 and 14 data-links, one SLQ-32(V)4 ESM system with intercept element, one SLQ-29(V)5 ECM system with WLR-8 warning and SLQ-17AV jamming elements, four Mk 36 Super RBOC chaff/flare launchers, one OE-82 satellite

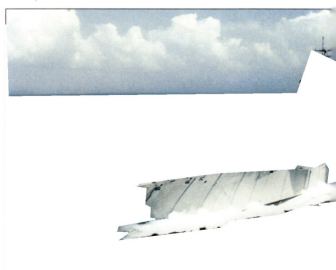

## ENTERPRISE CLASS

communication system, one SSR-1 satellite communication receiver, and URN-25 TACAN

**Propulsion:** Eight Westinghouse A2W pressurised water-cooled reactors supplying steam to four sets of Westinghouse geared turbines delivering 280,000hp (208,795kW) to four shafts

**Performance:** Maximum speed 35kt; range 460,000 miles (740,280km) at 20kt

**Complement:** 180+3,139 plus an air group of 310+2,010

| USA | | | |
|---|---|---|---|
| *Name* | *No.* | *Builder* | *Commissioned* |
| *Enterprise* | CVN65 | Newport News | Nov 1961 |

**Note**

The world's first nuclear-powered aircraft carrier, the *Enterprise* was laid down in 1958 and commissioned in November 1961 as the largest warship ever built (now exceeded in size by the aircraft carriers of the 'Nimitz'

*The* Enterprise *was the world's first nuclear-powered aircraft carrier and second nuclear-powered warship, the cruiser* Long Beach *having been completed a few months earlier, and is in essence a version of the 'Forrestal' class with a revised propulsion arrangement. The ship carries a large multi-role air group, and the flightdeck, with a substantial angled section, is served by four deck-edge lifts (three to starboard and one to port) as well as four C13 Mod 1 steam catapults (two each on the straight and angled sections of the flightdeck).*

# Modern Aircraft Carriers

*The* Enterprise *was completed without armament in an effort to minimise cost, but was later revised with three Sea Sparrow surface-to-air missile systems and subsequently with three Phalanx close-in weapon system mountings for last-ditch defence against anti-ship missiles. Considerable internal volume is occupied by the basic powerplant, which comprises eight reactors and 32 heat exchangers supplying steam to the four geared steam turbines that power the four shafts.*

class). The *Enterprise* was built to a modified 'Forrestal' class design with her larger dimensions dictated by the powerplant of eight Westinghouse A2W pressurised water-cooled nuclear reactors. The high cost of her construction prevented the subsequent implementation of the original plan for five sister ships.

From January 1979 to March 1982 the *Enterprise* was extensively refitted in a programme that included the rebuilding of her island superstructure and the fitting of new radar systems and mast to replace the characteristic ECM dome and billboard radar antennae. The *Enterprise* is equipped with four 295ft 0in (89.9m) steam catapults and four deck-edge aircraft lifts, and carries 2,520 tons of aviation ordnance plus 2.26 million Imp gal (10.3 million litres) of aircraft fuel to provide sufficiency for 12 days of sustained air operations before replenishment is required. Like that of other US carriers, the *Enterprise*'s ordnance load during the period of the Cold War included 10kT B61, 20kT B51, 60kT B43, 100kT B61, 200kT B43, 330kT B61, 400kT B43, 600kT B43 and 900kT B61 nuclear bombs, 100kT Walleye air-to-surface missiles and 10kT B51 depth bombs, while 1.4Mt B43 and 1.2Mt B28 strategic bombs could also be embarked as required. The air group is similar in size and configuration to that carried by the 'Nimitz' class carriers, and the *Enterprise* is fitted with the same command facilities. In addition to her OE-82 satellite communication system, since 1976 she has carried two British SCOT satellite communications antennae for use with British fleet units and NATO.

## FORRESTAL CLASS

# 'Forrestal' class

**Country of origin:** USA

**Type:** Conventionally powered multi-role aircraft carrier

**Displacement:** 59,060 tons or 60,000 tons (CV61/62) standard and 79,250 tons (CV59), 80,385 tons (CV60), 81,165 tons (CV61) or 80,645 tons (CV62) full load

**Dimensions:** Length 1,086ft 0in/331.0m (CV59) or 1,071ft 0in/326.4m (others); beam 129ft 6in (39.5m); draught 37ft 0in (11.3m); flightdeck length 1,047ft 0in (319.1m) and width 252ft 0in (76.8m)

**Gun armament:** Three 20mm Phalanx Mk 15 CIWS mountings

**Missile armament:** Three Mk 29 octuple launchers for RIM-7 NATO Sea Sparrow surface-to-air missiles

**Torpedo armament:** None

**Anti-submarine armament:** Aircraft and helicopters (see below)

*The first aircraft carrier designed and built after World War II, the Forrestal marked the emergence of a new concept in fleet aircraft-carrier design with its huge size and full optimisation for sustained flight operations through the provision of a multi-role air group using a two-section flightdeck equipped with four deck-edge lifts (three to starboard and one to port) and four steam catapults.*

# Modern Aircraft Carriers

**Aircraft:** Typically 90 in a multi-role carrier air wing with 20 Grumman F-14 Tomcat, 20 McDonnell Douglas F/A-18 Hornet, 20 Grumman A-6E Intruder, 4 Grumman KA-6D Intruder, 5 Grumman EA-6B Prowler, 10 Lockheed S-3A Viking and 5 Grumman E-2C Hawkeye fixed-wing aircraft, and 6 Sikorsky S-3G/H Sea King helicopters

**Electronics:** One SPS-48C 3D radar, one SPS-49(V)5 air-search radar, one SPS-67 surface-search radar, one SPS-58 low-level threat-warning radar, one SPN-64 navigation radar, four SPN-series carrier landing radars, one Mk 23 and four or six Mk 91 radars used in conjunction with two (CV59 and 62) or three (others) Mk 91 SAM fire-control systems, one NTDS, Links 4A, 11 and 14 data-links, one SLQ-29(V)3 ESM system with intercept element, one SLQ-26 ECM system with WLR-1, WLR-3 and WLR-11 warning elements, four Mk 36 Super RBOC chaff/flare launchers, one OE-82 satellite communication system, one WSC-3 satellite communication transceiver, one SSR-1 satellite communication receiver, and URN-25 TACAN

**Propulsion:** Eight Babcock & Wilcox boilers supplying

# FORRESTAL CLASS

*The Saratoga was completed as the second unit of the four-strong 'Forrestal' class.*

steam to four sets of Westinghouse geared turbines delivering 260,000hp/193,885kW (CV59) or 280,000hp/208,795kW (others), in each case to four shafts

**Performance:** Maximum speed 33kt (CV59) or 34kt (others); range 9,200 miles (14,805km) at 20kt or 4,600 miles (7,400km) at 30kt

**Complement:** 148+2,810 (CV59), 136+2,760 (CV60), 161+2,728 (CV61) or 150+2,643 (CV62) plus an air group of 290+2,190

### USA

| Name | No. | Builder | Commissioned |
|---|---|---|---|
| Forrestal | CV59 | Newport News | Oct 1955 |
| Saratoga | CV60 | New York NY | Apr 1956 |
| Ranger | CV61 | Newport News | Aug 1957 |
| Independence | CV62 | New York NY | Jan 1959 |

# MODERN AIRCRAFT CARRIERS

### Note
The four ships of the 'Forrestal' class were originally conceived as smaller versions of the ill-fated strategic aircraft-carrier design, the *United States*, with four aircraft catapults and a flush flightdeck fitted without an island. However, following a complete redesign the ships were completed as the first carriers designed and built specifically for jet aircraft operations, with a conventional island, an angled flightdeck to allow the four catapults to be retained, and four deck-edge elevators each measuring 72ft 0in (21.9m) by 52ft 0in (15.8m). The ships were initially designated as CVBs (heavy aircraft carriers), but were subsequently redesignated as CVAs (attack aircraft carriers) and finally as CVs (multi-purpose aircraft carriers), and the *Forrestal* has a less advanced propulsion arrangement with boilers working at lower pressure and at a higher fuel consumption.

# FORRESTAL CLASS

**Above:** *One of the most important warplanes carried by the 'Forrestal' class aircraft carriers between the late 1960s and late 1980s was the Vought A-7 Corsair II medium attack type.*

**Left:** *The Forrestal carries 5,500 tons of fuel for her air group, which uses a flightdeck accessed by four 99,000lb (44,906kg) lifts, fitted with four steam catapults for the launch of warplanes, and carrying four sets of arrester wires on the angled section for the recovery of aircraft.*

The aviation ordnance load of each ship is 1,650 tons; 624,515 Imp gal (2.839 million litres) of AVGAS aviation fuel and 157,380 Imp gal (595,740 litres) of JP5 aviation fuel are carried for the air wing embarked. The air group is similar to those of the larger carriers, and each ship has the standard four aircraft elevators serving the flightdeck. All the ships have undergone a Service Life Extension Program (SLEP) refit to extend their service lives to about the year 2000.

During the Grenada landings of November 1983 the *Constellation* provided the air cover and strike support to the US Marine Corps and US Army Ranger forces involved in the assault, while also maintaining ASW cover against any possible incursions by the two Cuban 'Foxtrot' class attack submarines. In order to rectify some of the deficiencies encountered in combat operations, the SLEP refits have improved habitability, added Kevlar armour to enclose the vital machinery and electronics spaces, improved the NTDS fitted, added a Task Force Command Center (TFCC) facility, and replaced the catapults. The radar outfit has also been upgraded and the air-defence armament was strengthened with the addition of the Phalanx 'Gatling' gun mounting for anti-missile use.

# MODERN AIRCRAFT CARRIERS

## 'Garibaldi' class

**Country of origin:** Italy

**Type:** Conventionally powered light anti-submarine aircraft carrier

**Displacement:** 10,100 tons standard and 13,370 tons full load

**Dimensions:** Length 590ft 7in (180.0m); beam 110ft 7in (33.4m); draught 22ft 0in (6.7m); flightdeck length 570ft 3in (173.8m) and width 99ft 9in (30.4m)

**Gun armament:** Six 40mm Breda L/70 AA in three Dardo twin mountings

**Missile armament:** Four Teseo container-launchers for 10 Otomat Mk 2 anti-ship missiles, and two Albatros octuple launchers for 48 Aspide surface-to-air missiles

**Torpedo armament:** None

**Anti-submarine armament:** Two triple B-515 tube mountings for 12.75in (324mm) A 244/S or Mk 46 torpedoes, and helicopter-launched weapons (see below)

**Aircraft:** Typically 16 Agusta (Sikorsky) ASH-3D Sea King helicopters, or 12 McDonnell Douglas AV-8B Harrier II STOVL warplanes, or a mix of the two types

**Electronics:** One SPS 52C 3D radar, one RAN 3L air-search radar, one RAN 10S air/surface-search radar, one SPS 702 UPX surface-search and target designation radar, one SPN 728 air-control radar, one SPN 749 navigation radar, three RTN 30X radars and three NA 30 optronic directors used in conjunction with three Argo NA30 Albatros fire-control systems, three RTN 20X radars used in conjunction with three Dardo fire-control systems, one IPN 20 (SADOC 2) action information system, Links 11 and 16 data-links, one Raytheon DE 1164 active search bow sonar, one Nettuno SLQ-732 ESM system with warning and jamming elements, one SLQ-25 Nixie towed torpedo-decoy system, two Breda SCLAR-D chaff/flare

## GARIBALDI CLASS

*Planned for Mediterranean operations, the Giuseppe Garibaldi was schemed as a helicopter carrier for the anti-submarine and escort roles, but emerged as a more capable multi-role type with provision for the McDonnell Douglas AV-8B Harrier II STOVL warplane.*

launchers, one satellite communication and navigation system, and SRN-15A TACAN

**Propulsion:** COGAG (COmbined Gas turbine Or Gas turbine) arrangement, with four Fiat/General Electric LM2500 gas turbines delivering 80,000hp (59,655kW) to two shafts

**Performance:** Maximum speed 30kt; range 8,080 miles (13,000km) at 20kt

**Complement:** 550 plus an air group of 230 and provision for a flag staff of 45

| Italy | | | |
|---|---|---|---|
| Name | No. | Builder | Commissioned |
| Giuseppe Garibaldi | C551 | Italcantieri | Jul 1985 |

### Note
Designed as a light carrier with gas turbine propulsion for the embarkation of a helicopter force, the *Giuseppe Garibaldi*

# MODERN AIRCRAFT CARRIERS

*The fact that the Giuseppe Garibaldi was designed for the operation of VTOL and STOVL aircraft is indicated by her straight flightdeck with its 'ski-jump' forward ramp, angled upward at 6.5° to help the take-off of heavily laden fixed-wing STOVL warplanes.*

was built with features suiting her for the carriage and operation of STOVL fighters. The flightdeck is straight, but is fitted with a 6.5° 'ski-jump' over its forward section to facilitate take-off by heavily laden STOVL warplanes. The hangar is 360ft 11in (110m) long, 49ft 3in (15.0m) wide and 19ft 8in (6.0m) high, and was built to accommodate 12 Agusta (Sikorsky) ASH-3 Sea King ASW helicopters or 10 STOVL aircraft and one Sea King, although the available height also permits the embarkation of heavy-lift Meridionali (Boeing Vertol) CH-47 Chinook helicopters if required. Two aircraft lifts are fitted (one forward and one abaft the island), and there are six marked flightdeck spaces for helicopter operations. The *Giuseppe Garibaldi* was designed for the specific task of providing ASW support for naval task forces and merchant convoys, and as such is fitted with full flagship facilities as well as command, control and communication systems for both naval and air force operations. In emergencies she can also carry up to 600 troops for short periods. The extensive weaponry also allows her to operate as an independent surface unit.

The ship is fitted with two pairs of fin stabilisers to permit helicopter operations in heavy weather, and her aircraft maintenance facilities are sufficient to service her own air group and the light ASW helicopters of any escorting warships. The ship can also carry two 250-man fast personnel launches for use in disaster relief or amphibious operations.

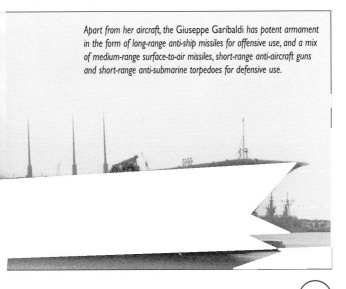

Apart from her aircraft, the Giuseppe Garibaldi *has potent armament in the form of long-range anti-ship missiles for offensive use, and a mix of medium-range surface-to-air missiles, short-range anti-aircraft guns and short-range anti-submarine torpedoes for defensive use.*

# MODERN AIRCRAFT CARRIERS

## 'Hermes' class

**Country of origin:** UK, now India

**Type:** Conventionally powered multi-role aircraft carrier

**Displacement:** 23,900 tons standard and 28,700 tons full load

**Dimensions:** Length 744ft 4in (226.9m); beam 90ft 0in (27.4m); draught 28ft 6in (8.7m); flightdeck length 744ft 4in (226.9m) and width 160ft 0in (48.8m)

**Gun armament:** Some 30mm ADGM-630 CIWS mountings may be installed

**Missile armament:** Two GWS 22 quadruple launchers for Sea Cat surface-to-air missiles, although these may be replaced by Russian surface-to-air missile systems

**Torpedo armament:** None

**Anti-submarine armament:** Helicopter-launched weapons (see below)

**Aircraft:** Typically 12 out of a theoretical strength of 30 BAe Sea Harrier FRS.Mk 51 STOVL fighters and 7 Westland Sea King Mk 42 helicopters

**Armour:** Belt 1–2in (25.4-50.8mm) over machinery spaces and magazines; deck 0.75in (19mm)

**Electronics:** One Type 996 air-search radar, one Type 994 air/surface-search radar, one Type 1006 navigation radar, Soviet fire-control radars to be fitted to complement the two Type 904 fire-control radars retained for the GWS 22 Sea Cat surface-to-air missile fire-control systems, one CAAIS action information system (perhaps to be replaced by a

# HERMES CLASS

Soviet system), Link 10 data-link, one Type 184M active search and attack hull sonar, one UAA-1 Abbey Hill ESM system with warning and jamming elements, two Corvus chaff launchers, and FT 13-S/M TACAN

**Propulsion:** Four Admiralty boilers supplying steam to two sets of Parsons geared turbines delivering 76,000hp (56,675kW) to two shafts

**Performance:** Maximum speed 28kt

**Complement:** 143+1,207 including air group

| India | | | |
|---|---|---|---|
| Name | No. | Builder | Commissioned |
| Viraat | R22 | Vickers | Nov 1959 |

**Note**
The original *Hermes* of the World War II period was designed and laid down as the sixth vessel of the 'Centaur' class, but in October 1945 the ship was cancelled and the

*After final service as the flagship of the task force that retook the Falkland Islands in 1982, the elderly British aircraft carrier* Hermes *was sold to India in May 1986 and became the Viraat.*

# Modern Aircraft Carriers

name given to the hull laid down as the *Elephant* of the same class. As very little work had been done on the original hull, the vessel was able to benefit from a complete redesign and was commissioned in November 1959 with a 6.5° angled flightdeck, a deck-edge aircraft lift as one of the two lifts fitted, and a 3D radar system. In 1964–66 the new *Hermes* was retrofitted with two quadruple Sea Cat surface-to-air missile systems in place of her original armament of ten 40mm Bofors AA guns in five twin mountings, and access to the seaward side of the island was provided.

In a further refit in 1971, the Type 984 3D radar was replaced by a Type 965 system, and a comprehensive deck-landing light system was fitted after the ship had been paid off for conversion to a commando assault carrier, as she could operate only a 28-aircraft group of de Havilland Sea Vixen, Blackburn Buccaneer and Fairey Gannet fixed-wing aircraft but not the modern McDonnell Douglas Phantom then entering service with the Fleet Air Arm.

During this conversion the *Hermes* also lost her arrester wires and catapult, and was converted to carry a complete Marine Commando unit with its associated squadron of Westland Wessex assault helicopters. By 1977 the *Hermes* was again being revised, in this instance to become an ASW carrier, although she retained the Commando-carrying ability. As such, she carried nine Westland Sea King ASW and four Wessex HU.Mk 5 utility helicopters. In 1980 the *Hermes* was taken in hand for her third major conversion, and yet again changed her role; in this instance a strengthening of the flightdeck and the provision of a 7.5° 'ski-jump' overhanging the bow allowed the operation of five BAe Sea Harriers in place of the Wessex machines.

In 1982, because of her extensive communications fit and greater aircraft-carrying ability, the *Hermes* was made flagship of the British task force that was sent to the Falklands. During this operation the aircraft carrier initially operated an air group of 12 Sea Harriers, nine Sea King HAS.Mk 5s and nine Sea King HC.Mk 4s. As the campaign progressed, however, the group was modified to 15 Sea Harriers, six Harrier GR.Mk 3s, five ASW Sea Kings, and two Westland Lynxes that were equipped for Exocet decoy operations. Following her success in the Falklands it was announced that the *Hermes*, after a series of deployments in 1983, would be refitted at the end of the year and then used in a training ship capacity in harbour because she was too 'labour intensive' and was not converted to use Dieso (the Royal Navy's current fuel type). The ship was sold to India in 1985.

# INVINCIBLE CLASS

# 'Invincible' class

**Country of origin:** UK

**Type:** Conventionally powered light multi-role aircraft carrier

**Displacement:** 19,500 tons or 20,000 tons (R07) standard

**Dimensions:** Length 677ft 0in (206.6m) or 685ft 10in/209.1m (R07); beam 90ft 0in (27.5m) and width 105ft 0in (31.9m) or 118ft 0in/36.0m (R07); draught 26ft 0in (7.9m); flightdeck length 550ft 0in (167.8m) and width 44ft 4in (13.5m) with fore and aft lateral extensions to the maximum width of each ship

**Gun armament:** Three 30mm Goalkeeper CIWS mountings, two 20mm Oerlikon AA in GAM-B01 single mountings, or two 30mm Oerlikon AA in one GCM-A03 twin mounting (R07)

**Missile armament:** One twin launcher for 36 Sea Dart surface-to-air missiles and four GWS 26 Mod 2 Lightweight Sea Wolf launchers for Sea Wolf surface-to-air missiles

*Like the Italian Giuseppe Garibaldi, the units of the 'Invincible' class are revealed as the operators of VTOL and STOVL warplanes by their flightdecks, which are straight, comparatively narrow, and are fitted at their forward ends with a 'ski-jump' ramp to facilitate the take-off of heavily laden STOVL warplanes.*

# Modern Aircraft Carriers

The Invincible, being refuelled at sea by the fleet oil tanker Olmeda, is the first unit of a three-ship class named after her and designed as a 'through-deck cruiser' but in fact completed as the small but effective platform for VTOL and STOVL aircraft such as the Westland Sea King anti-submarine helicopter and the BAe Sea Harrier multi-role STOVL warplane.

**Torpedo armament:** None

**Anti-submarine armament:** Helicopter-launched weapons (see below)

**Aircraft:** Since 1988 the ships have been upgraded to CVSA (carrier vessel submarine attack) configuration with a normal air group of 12 Westland Sea King helicopters (3 AEW.Mk 2s and 9 HAS.Mk 5s) and 9 BAe Sea Harrier FRS.Mk 1/2 STOVL fixed-wing aircraft, with the possibility that European Helicopter Industries EH.101 Merlin helicopters may replace the Sea Kings during the later 1990s

**Electronics:** One Type 1022 air-search radar, one Type 992R surface-search radar, two Type 1006 navigation radars, two Type 909 radars used in conjunction with two GWS 30 Sea Dart SAM fire-control systems, two Type

## INVINCIBLE CLASS

911(3) radars used in conjunction with four GWS 26 Mod 2 fire-control systems (when Sea Wolf surface-to-air missiles are retrofitted), one ADAWS 6 or ADAWS 10 (R07) action information system, Link 10, 11 and 14 data-links, one Type 2016 active search and attack hull sonar (R07 only), one UAA-1 Abbey Hill ESM system with warning and jamming elements, two Corvus chaff launchers, two Mk 36 Super RBOC chaff/flare launchers, two Sea Gnat chaff launchers (R07 only), one OE-82 satellite communication system, one SCOT-2 satellite communication system, and one SCOT-1 satellite communication system

**Propulsion:** COGAG (COmbined Gas turbine And Gas turbine) arrangement with four Rolls-Royce Olympus TM3B gas turbines delivering 94,000hp (70,085kW) to two shafts

**Performance:** Maximum speed 28kt; range 5,750 miles (9,250km) at 18kt

**Complement:** 57+609 plus an air group of 84+318, the post-1988 update adding a further 120 personnel

## MODERN AIRCRAFT CARRIERS

### UK

| Name | No. | Builder | Commissioned |
|------|-----|---------|--------------|
| Invincible | R05 | Vickers | Jul 1980 |
| Illustrious | R06 | Swan Hunter | Jun 1982 |
| Ark Royal | R07 | Swan Hunter | Jul 1985 |

**Note**
The death of the British fixed-wing aircraft carrier, when the CVA01 fleet carrier replacement programme was

## INVINCIBLE CLASS

*Unlike many other British warships of the present day, ships of the 'Invincible' class (left and above) have an effective anti-aircraft capability in the Sea Dart long-range missile system, Sea Wolf medium/short-range missile system, and Goalkeeper close-in weapon system with its radar-controlled 30mm multi-barrel cannon.*

terminated in 1966, led to the issue of a Staff Requirement for a 12,500-ton command cruiser equipped for the carriage of six Westland Sea King ASW helicopters. A revision of this basic concept as well as a number of operational studies showed that a nine-helicopter air group would be more effective. These new specifications resulted in a design draft which became known as the 19,500-ton 'through-deck cruiser' (TDC) design, the term TDC being used for what was essentially a light carrier design because of the political sensitivity with which politicians viewed the possibility of a carrier resurrection at the time. Despite this, the designers showed initiative in allowing sufficient space

# Modern Aircraft Carriers

*The* Invincible *(above) and the* Illustrious *and* Ark Royal *(right) provide the Royal Navy with the possibility of having two aircraft carriers always ready for service, a fact that would have been impossible if the 1982 scheme to sell one ship to Australia had gone ahead.*

and facilities to be incorporated for a naval version of the RAF's current STOVL close-support fighter programme. The designers were duly awarded for such foresight in May 1975 when it was announced officially that the TDC would carry the BAe Sea Harrier.

Thus the *Invincible*, first ship of the new class, was not delayed during building. In May 1916 the second ship was ordered as the *Illustrious*, and in December 1918 a contract was placed for the third unit, the *Indomitable*. As a result of public displeasure caused by the paying off of the last fixed-wing carrier, *Ark Royal*, the Admiralty decided to change the name of the third ship to *Ark Royal* in order to placate public opinion.

At the time of their completion, the ships of this 'Invincible' class were the largest gas turbine-powered warships in the world, with virtually every piece of equipment below decks, including the engine modules, and suitable for maintenance by exchange. During construction, both the *Invincible* and the *Illustrious* were fitted with 7° 'ski-jumps', while the *Ark Royal* was completed with a 12° 'ski-jump' of the type later retrofitted on the first two ships.

In February 1982, a surprising defence statement announced that the *Invincible* was to be sold to Australia as a helicopter carrier to replace the *Melbourne*, leaving only two of the carriers in British service. This deal was cancelled after the Falklands campaign of 1982, however: to the great

## INVINCIBLE CLASS

relief of the naval high command, the government had finally realised that three carriers had to be available to ensure that two were in service at any one time.

During Operation 'Corporate', as the Falklands campaign was codenamed, the *Invincible* started with an air group of eight Sea Harriers and nine Sea King HAS.Mk 5 ASW helicopters. As a result of losses and replacements, however, this was modified to a group of 11 Sea Harriers, eight ASW Sea Kings, and two Lynx helicopters configured for Exocet decoy duties: most of the extra aircraft had to be accommodated on the deck as there was insufficient room for them in the hangar. The *Illustrious* was hurried through to completion in time to relieve the *Invincible* after the war, and she sailed south with 10 Sea Harriers, nine ASW Sea Kings and two Sea King AEW conversions. She and her sister ships were also fitted with two American 20mm Phalanx 'Gatling' CIWS mountings for anti-missile defence and two single 20mm AA guns to improve the close-in air defences.

Until recently, all three carriers carried nuclear weapons for their air groups. According to official sources, the Sea Kings can carry American nuclear depth bombs and the Sea Harriers are believed to be capable of carrying tactical nuclear bombs. Although conceived originally as anti-submarine ships, these vessels are now useful multi-role types as a result of an embarked complement of Sea Harrier STOVL aircraft. Considerable effort has been made to improve the close-range defence of these ships with more-capable SAM and CIWS systems, and enhanced decoy measures, and during retrofits accommodation is being provided for an additional 120 aircrew and flag staff.

# MODERN AIRCRAFT CARRIERS

## 'John F. Kennedy' class

**Country of origin:** USA

**Type:** Conventionally powered multi-role aircraft carrier

**Displacement:** 61,000 tons standard and 80,940 tons full load

**Dimensions:** Length 1,052ft 0in (328.7m); beam 130ft 0in (39.6m); draught 35ft 11in (10.9m); flightdeck length 1,052ft 0in (328.7m) and width 267ft 6in (81.6m)

**Gun armament:** Three 20mm Phalanx Mk 15 CIWS mountings

**Missile armament:** Three Mk 29 octuple launchers for RIM-7 NATO Sea Sparrow surface-to-air missiles

# JOHN F. KENNEDY CLASS

**Torpedo armament:** None

**Anti-submarine armament:** Aircraft and helicopters (see below)

**Aircraft:** Typically 90 in a multi-role carrier air wing with 20 Grumman F-14 Tomcat, 20 McDonnell Douglas F/A-18 Hornet, 20 Grumman A-6E Intruder, 5 Grumman EA-6B Prowler, 4 Grumman KA-6D Intruder, 5 Grumman E-2C Hawkeye and 10 Lockheed S-3A Viking fixed-wing aircraft, and 6 Sikorsky SH-3G/H Sea King helicopters

**Electronics:** One SPS-48C/E 3D radar, one SPS-49 long-range air-search radar, one SPS-65 low-level threat-warning

*This overhead view of the John F. Kennedy reveals the enormous size of the flightdeck with its straight and angled sections, four deck-edge lifts (three to starboard and one to port), and part of the very large and mixed warplane complement.*

## MODERN AIRCRAFT CARRIERS

radar, one SPS-10F surface-search radar, one SPN-64 navigation radar, four SPN-series carrier landing radars, or Mk 23 and six Mk 57 radars used in conjunction with three Mk 91 SAM fire-control systems, one NTDS, Links 11 and 14 data-links, one SLQ-17 and one SLQ-26 ESM systems with WLR-3 and WLR-11 warning and jamming elements, four Mk 36 Super RBOC chaff/flare launchers, one OE-82 satellite communication system, one WSC-3 satellite communication transceiver, one SSR-1 satellite communication receiver, and URN-25 TACAN

**Propulsion:** Eight Foster-Wheeler boilers supplying steam to four sets of Westinghouse geared turbines delivering 280,000hp (208,795kW) to four shafts

**Performance:** Maximum speed 32kt; range 9,200 miles (14,805km) at 20kt

**Complement:** 155+3,045 plus an air group of 320+2,160

| USA | | | |
| --- | --- | --- | --- |
| Name | No. | Builder | Commissioned |
| *John F. Kennedy* | CV67 | Newport News | Sep 1968 |

# JOHN F. KENNEDY CLASS

*The John F. Kennedy was completed to an improved 'Forrestal' class design, and the flightdeck includes one C13 Mod 1 and three C13 steam catapults (two each on the straight and angled sections) as well as four arrester wires on the angled section.*

### Note
This is a close relative of the three 'Kitty Hawk' class aircraft carriers, all four ships being variants of an improved 'Forrestal' class design with the island set farther aft. The ship has an angled flightdeck served by four steam catapults and four deck-edge lifts.

# Modern Aircraft Carriers

## 'Kiev' class

**Country of origin:** USSR, now CIS

**Type:** Conventionally powered multi-role hybrid aircraft carrier/guided-missile cruiser

**Displacement:** 40,500 tons full load

**Dimensions:** Length 899ft 0in (274.0m); beam 107ft 3in (32.7m) and width 154ft 10in (47.2m); draught 32ft 10in (10.0m); flightdeck length 620ft 0in (189.0m) and width 68ft 0in (20.7m) with fore and aft lateral extensions to the maximum width of each ship

**Gun armament:** Four 3in (76mm) L/60 DP in two twin mountings, and eight 30mm ADGM-630 CIWS mountings

**Missile armament:** Four twin container-launchers for 24 SS-N-12 'Sandbox' anti-ship missiles, two twin launchers for 72 SA-N-3B 'Goblet' surface-to-air missiles, two twin launchers for about 40 SA-N-4 'Gecko' surface-to-air missiles or two groups of six octuple vertical launchers for 96 SA-N-9 surface-to-air missiles (*Novorossiysk* only)

# Kiev Class

**Torpedo armament:** Two 21in (533mm) quintuple tube mountings for Type 53 dual-role torpedoes

**Anti-submarine armament:** One twin SUW-N-1 launcher for 20 FRAS-1 (free rocket anti-submarine)/SS-N-14 'Silex' missiles, two RBU 6000 12-barrel rocket-launchers, torpedoes (see above), and helicopter-launched weapons (see below)

**Aircraft:** Typically 32, comprising 13 fixed-wing (12 Yakovlev Yak-38 'Forger-A' and one Yak-38 'Forger-B') machines and 19 rotary-wing (16 Kamov Ka-27 'Helix-A' and 3 Ka-27 'Helix-B') machines

**Electronics:** One 'Top Sail' 3D air-surveillance radar, one 'Top Steer/Top Plate' 3D air/surface radar, two 'Strut Pair' air-search radars (*Novorossiysk* only), one 'Trap Door' anti-ship missile fire-control radar, two 'Head Light-C' SA-N-3 fire-control radars, two 'Pop Group' SA-N-4 or four 'Cross Sword' SA-N-9 (*Novorossiysk*) fire-control radars, two 'Owl Screech' 3in (76mm) gun fire-control radars, four

*The Kiev was the lead unit of a three-ship class designed to pave the way for the development of the USSR's navy, with a genuine 'blue water' capability through the provision of organic air power for widely deployed task forces.*

49

# Modern Aircraft Carriers

'Bass Tilt' CIWS fire-control radars, one 'Top Knot' aircraft-control radar, four 'Tee Plinth' or 'Tin Man' (*Novorossiysk*) optronic trackers, one 'Don Kay' and two 'Palm Frond' or three 'Palm Frond' (*Novorossiysk*) navigation radars, one 'Shot Dome' navigation radar, one low-frequency active search and attack hull sonar, one medium-frequency active search variable-depth sonar, one 'High Pole-A' and one 'High Pole-B' or one 'Salt Pot-A', one 'Salt Pot-B' and one 'Square Head' (*Novorossiysk*) IFF systems, an extensive electronic warfare suite including

# KIEV CLASS

*Third unit of the 'Kiev' class, the Novorossiysk was typical of Soviet warships in her attractive lines and heavy but well-balanced offensive/defensive armament complemented by a mass of electronic systems.*

eight 'Bell Globe' (not in *Novorossiysk*), four 'Rum Tub', two 'Bell Bash', four 'Bell Nip' and two 'Cage Pot' (not in *Novorossiysk*) antennae/housings, two twin chaff/flare launchers, one towed torpedo-decoy system, and a large variety of conventional and satellite communication and navigation systems including a 'Punch Bowl' antenna for SS-N-12 midcourse guidance update via satellite

# Modern Aircraft Carriers

**Propulsion:** Eight boilers supplying steam to four sets of geared turbines delivering 201,180hp (150,025kW) to four shafts

**Performance:** Maximum speed 32kt; range 14,975 miles (24,095km) at 18kt or 4,600 miles (7,400km) at 30kt

**Complement:** 1,200 excluding air group

### USSR, now CIS

| Name | Builder | Commissioned |
|---|---|---|
| Kiev | Nikolayev South | May 1975 |
| Minsk | Nikolayev South | Feb 1978 |
| Novorossiysk | Nikolayev South | Aug 1982 |

**Note**
Planning for a class of hybrid cruiser/carriers probably started in the USSR during the early 1960s, when the need was perceived for a small number of ships able to carry the interceptor aircraft that could provide air defence for Soviet submarine-hunting forces and for submarines operating in hostile sea areas. Although the 'Moskva' class of helicopter cruisers was meant to be in series production, with a total of 12 such ships planned, the units of the class could operate only ASW helicopters, and so did not meet the new requirement. Thus the yard building the 'Moskva' class, Black Sea Shipyard No. 444 at Nikolayev, switched to construction of a new design after the completion of only two 'Moskva' class ships.

The first of the new four-ship class, the *Kiev*, was commissioned in May 1975 after extensive trials. The second ship, the *Minsk*, followed in February 1978, and the third ship, the *Novorossiysk*, was completed to a modified design in August 1982. A fourth vessel, the *Baku*, was completed to a somewhat modified design and later became the *Admiral Gorshkov*.

The *Kiev* and her sister ships are basically the same in physical appearance but differ in equipment fits. All have seven take-off and landing spots marked on their angled flightdecks on the port side: six (marked with the letter C and the numbers 1 to 6) are for helicopters, whilst the Yakovlev Yak-38 'Forger-A' STOVL fighters take off from

# KIEV CLASS

*Second unit of the 'Kiev' class to be completed, the Minsk reveals the three-quarter-length and slightly angled flightdeck of the class, which was designed to operate helicopters and STOVL warplanes.*

position 6 and land on the asbestos tile-coated flightdeck near position 5 on a specially designated spot (marked with an E on the *Kiev* and with an M on the *Minsk* and the *Novorossiysk*). Each of the first two ships has seven deck lifts: a 23ft 0in (7.0m) square unit for cargo on the port side of the island forward; a 30ft 2in (9.2m) by 34ft 0in (10.35m) unit for helicopters close to the island midpoint; a 60ft 8in (18.5m) by 15ft 5in (4.7m) unit aft of the island for the 'Forgers'; three 21ft 4in (6.5m) by 5ft 0in (1.5m) weapons lifts in line astern next to the helicopter lift; and a similarly sized personnel lift on the port side. The third ship does not have the cargo lift and has only two ammunition lifts; in addition, she has a test area for running-up the 'Forger' engines on the starboard side of the flight deck.

The air wing for the 'Kiev' class initially comprised 12

# Modern Aircraft Carriers

*The Kiev reveals the substantial helicopter and fixed-wing warplane parking area to starboard of the flightdeck and abaft the superstructure.*

Yak-38s (including one 'Forger-B' unarmed conversion trainer), 18 Kamov Ka-25 'Hormone-A' ASW helicopters, three 'Hormone-B' midcourse missile-guidance correction/-target designator/ELINT helicopters and one 'Hormone-C' SAR planeguard helicopter. All the embarked aircraft can be carried in the hangar.

The *Novorossiysk* is fitted with the SA-N-9 surface-to-air missile system instead of the SA-N-4 'Gecko' system carried by her sisters. As far as electronics are concerned, the *Novorossiysk* does not carry the characteristic 'Side Globe' ESM domes on each side of the island superstructure, but instead has two as yet unidentified radars and four 'Bell Crown' optronic targeting systems that were first seen on the nuclear-powered battle-cruiser *Kirov*. In overall terms, these highly impressive ships are the largest combatants currently in full service with the CIS, and in their time marked a new departure from Soviet norms by providing an organic air capability in Soviet deep-water battle groups. The type possesses a high level of capability in several roles, but is also being used for the development of the various systems required for a fully fledged aircraft-carrier type able to operate conventional rather than STOVL aircraft. The sensor and weapon fits have been considerably improved in the third vessel, which is a hybrid type combining features of the 'Kiev' and 'Kiev (Modified)' classes. The flightdeck is angled at 4.5° and served by two lifts. The Soviets initially classified the type as an anti-submarine cruiser, since modified to a tactical aircraft-carrying cruiser.

# 'Kiev (Modified)' class

**Country of origin:** USSR, now CIS

**Type:** Conventionally powered multi-role hybrid aircraft carrier/guided-missile cruiser

**Displacement:** 40,500 tons full load

**Dimensions:** Length 899ft 0in (274.0m); beam 107ft 3in (32.7m) and width 154ft 10in (47.2m); draught 32ft 10in (10.0m); flightdeck length 620ft 0in (189.0m) and width 68ft 0in (20.7m) with fore and aft lateral extensions to the maximum width of each ship

**Gun armament:** Two 3.94in (100mm) L/70 DP in single mountings, and eight 30mm ADGM-630 CIWS mountings

**Missile armament:** Six twin container-launchers for 28 SS-N-12 'Sandbox' anti-ship missiles, and four groups of six octuple vertical launchers for 192 SA-N-9 surface-to-air missiles

**Torpedo armament:** None

**Anti-submarine armament:** Two new RBU-series 10-barrel rocket-launchers, and helicopter-launched weapons (see below)

**Aircraft:** Typically 32, comprising 13 fixed-wing (12 Yakovlev Yak-38 'Forger-A' and one Yak-38 'Forger-B') machines and 19 rotary-wing (16 Kamov Ka-27 'Helix-A' and 3 Ka-27 'Helix-B') machines

**Electronics:** One 'Sky Watch' 3D surveillance radar with four planar arrays, one 'Cylinder Blanc' air-search radar, one 'Top Steer/Top Plate' 3D air/surface radar, two 'Strut Pair' air-search radars, one 'Trap Door' anti-ship missile fire-control radar, four 'Cross Sword' surface-to-air missile fire-control radars, two 'Hawk Screech' 3.94in (100mm) gun fire-control radars, four 'Bass Tilt' CIWS fire-control radars, one 'Top Knot' aircraft-control radar, three 'Tin Man' optronic trackers, three 'Palm Frond' navigation radars, one 'Shot Dome' navigation radar, one low-frequency active

# Modern Aircraft Carriers

*The Baku is a development of the 'Kiev' class with modified missile armament.*

search and attack hull sonar, one medium-frequency active search variable-depth sonar, one 'Salt Pot-A', one 'Salt Pot-B' and one 'Cake Stand' IFF systems, an extensive electronic warfare suite including four 'Wine Flask', 12 'Bell' series and four 'Ball' series antennae/housings, two twin chaff/flare launchers, one towed torpedo-decoy system, and a large variety of conventional and satellite communication and navigation systems including two 'Punch Bowl' antennae for SS-N-12 midcourse guidance update via satellite and two 'Low Ball' satellite navigation systems

**Propulsion:** Eight boilers supplying steam to four sets of geared turbines delivering 201,180hp (150,025kW) to four shafts

**Performance:** Maximum speed 32kt; range 14,975 miles (24,095km) at 18kt or 4,600 miles (7,400km) at 30kt

**Complement:** 1,200 excluding air group

| USSR, now CIS | | |
|---|---|---|
| Name | Builder | Commissioned |
| Admiral Gorshkov | Nikolayev South | Jun 1987 |

**Note**

This ship, previously named the *Baku*, is a considerable improvement on the units of the 'Kiev' class, the most obvious modifications being improved electronics, revised gun and missile armament, alteration of the anti-submarine weapons, and deletion of the torpedo armament.

## KITTY HAWK CLASS

# 'Kitty Hawk' class

**Country of origin:** USA

**Type:** Conventionally powered multi-role aircraft carrier

**Displacement:** 60,100 tons standard and 81,125 tons full load (CV63), or 60,100 tons standard and 81,775 tons full load (CV64), or 60,300 tons standard and 79,725 tons full load (CV66)

**Dimensions:** Length 1,062ft 6in (324.0m) or 1,072ft 6in/327.1m (CV64) or 1,047ft 6in/319.5m (CV66); beam 129ft 6in (39.5m) or 130ft 0in/39.6m (CV66); draught 37ft 0in (11.3m); flightdeck length 1,062ft 6in (324.0m) or 1,072ft 6in/327.1m (CV64) or 1,047ft 6in/319.5m (CV66) and width 250ft 0in (76.2m) or 266ft 0in/81.1m (CV66)

**Gun armament:** Three 20mm Phalanx Mk 15 CIWS mountings

**Missile armament:** Three Mk 29 octuple launchers for RIM-7 NATO Sea Sparrow surface-to-air missiles

**Torpedo armament:** None

**Anti-submarine armament:** Aircraft and helicopters (see below)

**Aircraft:** Typically 90 in a multi-role carrier air wing with 20 Grumman F-14 Tomcat, 20 McDonnell Douglas F/A-18 Hornet, 20 Grumman A-6E Intruder, 5 Grumman EA-6B Prowler, 4 Grumman KA-6D Intruder, 5 Grumman E-2C Hawkeye and 10 Lockheed S-3A Viking fixed-wing aircraft, and 6 Sikorsky SH-3G/H Sea King helicopters

**Electronics:** One SPS-48C/E 3D radar, one SPS-49(V)5 long-range air-search radar, one SPS-65 low-level threat-warning radar, one SPS-10F or SPS-67 surface-search radar, one SPN-64(V)9 navigation radar, five SPN series carrier landing radars, one Mk 23 and two or three (CV66) Mk 57 radars used in conjunction with two or three (CV66) Mk 91 surface-to-air missile fire-control systems, one NTDS, Links 4A, 11 and 14 data-links, one SLQ-32(V) in

## Modern Aircraft Carriers

*The Kitty Hawk is the lead vessel of a three-ship class to which the later John F. Kennedy is a sister ship. The four ships are basically similar, and introduced the now-standard arrangement of four deck-edge lifts (one to port and three to starboard, the latter arranged as two forward and one abaft the island) that greatly improves the layout and operability of the flightdeck.*

succession to one SLQ-29(V)3 ESM system with WLR-8 warning and SLQ-17 jamming elements, four Mk 36 Super RBOC chaff/flare launchers, one OE-82 satellite communication system, one WSC-3 satellite communication transceiver, one SSR-1 satellite communication receiver, and URN-25 TACAN

**Propulsion:** Eight Foster-Wheeler boilers supplying steam to four sets of Westinghouse geared turbines delivering 280,000hp (208,795kW) to four shafts

**Performance:** Maximum speed 32kt; range 9,200 miles (14,805km) at 20kt

**Complement:** 139+2,634 (CV63), 156+2,861 (CV64), or 152+2,811 (CV66) plus an air group of 320+2,160

### USA

| Name | No. | Builder | Commissioned |
|---|---|---|---|
| Kitty Hawk | CV63 | New York SB | Apr 1961 |
| Constellation | CV64 | New York NY | Oct 1961 |
| America | CV66 | Newport News | Jan 1965 |

**Note**
Built to an improved 'Forrestal' class design, these three carriers in reality constitute two subclasses that are readily distinguishable from their predecessors by the fact that their island superstructures are set farther aft. In addition, two of their four aircraft lifts are forward of the island, the 'Forrestals' having only one in this location. A lattice radar mast is also carried abaft the island. The America is very similar to the Kitty Hawk and the Constellation, and was built in preference to an

# Kitty Hawk Class

austere nuclear-powered carrier; she is, however, the only US carrier of post-war construction to be fitted with a sonar system. Another unit, the *John F. Kennedy*, was built to a revised design incorporating an underwater protection system developed originally for the nuclear carrier programme. Each of the four ships has four steam catapults and carries some 2,150 tons of aviation ordnance plus 1.62 million Imp gal (7.38 million litres) of aviation fuel for their air groups.

The air groups are similar in size and composition to those embarked on the 'Nimitz' class carriers. The tactical reconnaissance element in each of the air wings is provided by three Grumman F-14s equipped with a tactical-aircraft reconnaissance pod system (TARPS).

The ships have the full ASW Classification and Analysis Center (ASCAC), TFCC and NTDS facilities (the *America* being the first carrier to be fitted with the NTDS), as well as the OE-82 satellite communication system, and were the first carriers able to undertake the simultaneous launch and recovery of aircraft without the type of problems that were encountered on earlier carriers. The ships were cycled through the SLEP from 1985 in order to extend their lives into the twenty-first century. Like the *John F. Kennedy*, the ships have also been fitted with a system to protect the magazines from ingress by sea-skimming anti-ship missiles.

# MODERN AIRCRAFT CARRIERS

## 'Kuznetsov' class

**Country of origin:** USSR, now CIS

**Type:** Conventionally powered multi-role aircraft carrier

**Displacement:** 67,500 tons full load

**Dimensions:** Length 999ft 0in (304.5m); beam 121ft 5in (37.0m); draught 34ft 5in (10.5m); flightdeck length 999ft 0in (304.5m) and width 229ft 8in (70.0m)

**Gun armament:** Six 30mm ADGM-630 CIWS mountings and eight CADS-N-1 hybrid 30mm cannon/SA-19 missile CIWS systems

**Missile armament:** 12 flush-fitting container-launchers for 12 SS-N-19 'Shipwreck' anti-ship missiles, and four groups of six octuple vertical launchers for 192 SA-N-9 surface-to-air missiles, and SA-N-11 surface-to-air missiles for the CADS-N-1 launchers (see above)

**Torpedo armament:** None

**Anti-submarine armament:** Two RBU-1200 rocket-launchers, and helicopter-launched weapons (see below)

**Aircraft:** About 24 fixed-wing (12 Sukhoi Su-27 'Flanker-B' and 12 Yakovlev Yak-38 'Forger-A') machines and 18 rotary-wing (Kamov Ka-27 'Helix') machines

**Electronics:** One 'Sky Watch' 3D surveillance radar with four planar arrays, one 'Top Plate' air-search radar, two 'Strut Pair' surface-search radars, three 'Palm Frond' navigation radars, four 'Bass Tilt', 'Kite Screech' or 'Owl Screech' gun fire-control radars, four 'Cross Sword' surface-to-air missile fire-control radars, four 'Tin Man' optronic trackers, two 'Punch Bowl' anti-ship missile guidance radars, one 'Fly Trap-B' aircraft-control radar, one 'Cake Stand' aircraft-control radar, data-link system, two 'Low Ball' satellite navigation systems, one 'Horse Jaw' hull-mounted search and attack sonar, four 'Wine Flask' and/or 'Bell' series ESM antennae/housings, eight 'Foot Ball' ESM antennae/housings, chaff/flare launchers, and other systems as yet unrevealed

# KUZNETSOV CLASS

*The Admiral Kuznetsov was the first Soviet aircraft carrier designed for the operation of an air group including conventional fixed-wing warplanes, as indicated by the angled flightdeck, although catapults would have to be retrofitted for exploitation of such a capability.*

**Propulsion:** Eight boilers supplying steam to four sets of geared turbines delivering 201,180hp (150,025kW) to four shafts

**Performance:** Maximum speed 32kt

**Complement:** 2,100 including air group

| USSR, now CIS | | |
| --- | --- | --- |
| Name | Builder | Commissioned |
| Admiral Kuznetsov | Nikolayev South | Jan 1991 |
| Varyag | Nikolayev South | Dec 1985 |

**Note**
The *Admiral Kuznetsov* and the *Varyag* (originally named *Tbilisi* and *Riga* respectively), were due to enter service in the early 1990s to provide Soviet (now Russian) deep-water battle groups with a true organic air component using conventional rather than STOVL aircraft. The flightdeck has a standard angled section, a forward 12° 'ski-jump', and three lifts (two on the starboard edge of the deck and the third inboard of the island); the second ship has an unknown number of steam catapults. Presently, it is not known which aircraft the carriers will embark in their air groups, though it is likely that both ships will carry Su-27s and Yak-38s.

# Modern Aircraft Carriers

## 'Majestic' class

**Country of origin:** UK, now India

**Type:** Conventionally powered light aircraft carrier

**Displacement:** 16,000 tons standard and 19,500 tons full load

**Dimensions:** Length 700ft 0in (213.4m); beam 80ft 0in (24.4m) and width 128ft 0in (39.0m); draught 24ft 0in (7.3m); flightdeck length 700ft 0in (213.4m) and width 112ft 0in (34.1m)

**Gun armament:** Seven 40mm Bofors L/70 AA in single mountings

**Missile armament:** None

**Torpedo armament:** None

**Anti-submarine armament:** Aircraft and helicopters (see below)

## MAJESTIC CLASS

**Aircraft:** Six fixed-wing (BAe Sea Harrier FRS.Mk 51) machines and 9 rotary-wing (Westland Sea King Mk 42) machines

**Electronics:** One LW-08 air-search radar, one DA-05 air/surface-search radar, one ZW-06 surface-search and navigation radar, one Type 963 aircraft landing radar, one IPN 10 action information system, and various communication and navigation systems

**Propulsion:** Four Admiralty boilers supplying steam to two sets of Parsons geared turbines delivering 40,000hp (29,830kW) to two shafts

**Performance:** Maximum speed 24.5kt; range 13,800 miles (22,210km) at 14kt or 7,150 miles (11,505km) at 23kt

**Complement:** 1,075 in time of peace and 1,345 in time of war

*Bought from the UK by India whilst still incomplete, the Vikrant has served for more than 35 years, in the process being transformed from a conventional light aircraft carrier into a platform for VTOL and STOVL aircraft.*

# Modern Aircraft Carriers

## India

| Name | No. | Builder | Commissioned |
|---|---|---|---|
| *Vikrant* | R11 | Vickers-Armstrongs | Mar 1961 |

**Note**

Formerly the British 'Glory' class carrier *Hercules* that had been laid up in an incomplete state since May 1946, the *Vikrant* was bought by India in January 1951, and was taken in hand by the Harland & Wolff yard in Belfast during April 1951 for completion with a single hangar, two electrically operated aircraft lifts, an angled flightdeck and a steam catapult. She was also partially fitted with an air-conditioning system for tropical service, and commissioned in March 1961. During the Indo-Pakistan war of 1971 the *Vikrant* operated a mixed air group of 16 Hawker Sea Hawk fighter-bombers and four Breguet Alizé ASW aircraft off East Pakistan (now Bangladesh), the elderly Sea Hawks attacking many coastal ports, airfields and small craft in a successful operation to prevent the movement of Pakistani men and supplies during Indian army operations to 'liberate' that country.

The *Vikrant* underwent a major refit between January 1979 and January 1982 to enable her to operate BAe Sea Harrier FRS.Mk 1 STOVL warplanes. Included in the refit was the construction of a 'ski-jump' ramp, the installation of new boilers and engines, the provision of new Dutch radars, and the fitting of a new operations control system. It is reported that the steam catapult has been retained in order for the carrier to continue operation of the Alizé aircraft, a number of which were refurbished at the same time for effective service up to the mid-1990s.

# 'Midway' class

**Country of origin:** USA

**Type:** Conventionally powered multi-role aircraft carrier

**Displacement:** 51,000 tons standard and 64,000 tons full load

**Dimensions:** Length 979ft 0in (298.4m); beam 121ft 0in (36.9m); draught 35ft 3in (10.8m); flightdeck length 979ft 0in (298.4m) and width 258ft 6in (78.8m)

**Gun armament:** Three 20mm Phalanx Mk 15 CIWS mountings

**Missile armament:** Two Mk 25 octuple launchers for RIM-7 Sea Sparrow surface-to-air missiles

**Torpedo armament:** None

**Anti-submarine armament:** Aircraft and helicopters (see below)

**Aircraft:** Typically 66 in a multi-role carrier air wing with 36 McDonnell Douglas F/A-18 Hornet, 12 Grumman A-6E Intruder or Vought A-7E Corsair II, 4 Grumman KA-6D Intruder, 4 Grumman EA-6B Prowler and 4 Grumman E-2C Hawkeye fixed-wing aircraft, and 6 Sikorsky SH-3H Sea King helicopters

**Electronics:** One SPS-48C/E 3D radar, one SPS-49 long-range air-search radar, one SPS-67 surface-search radar, one SPS-64 navigation radar, three SPN-series aircraft landing radars, two Mk 57 radars used in conjunction with two Mk 115 SAM fire control systems, one NTDS, Links 11 and 14 data-links, one SLQ-17(V) or SLQ-29 ESM system with WLR-1, WLR-10 and WLR-11 warning elements and ULQ-6 jamming element, four Mk 36 Super RBOC chaff/flare launchers, one OE-82 satellite communication system, one WSC-3 satellite communication transceiver, one SSR-1 satellite communication receiver, and URN-25 TACAN

# MODERN AIRCRAFT CARRIERS

*Commissioned one month after World War II's end, the Midway was the last relic of a previous age.*

**Propulsion:** 12 Babcock & Wilcox boilers supplying steam to four sets of Westinghouse geared turbines delivering 212,000hp (158,090kW) to four shafts

**Performance:** Maximum speed 30kt; range 17,275 miles (27,800km) at 15kt

**Complement:** 142+2,684 plus an air group of 1,854

| USA | | | |
| --- | --- | --- | --- |
| Name | No. | Builder | Commissioned |
| Midway | CV41 | Newport News | Sep 1945 |

**Note**
Originally a class of three, the *Midway*, the *Franklin D. Roosevelt* and the *Coral Sea* (commissioned in September 1945, October 1945 and October 1947 respectively) were the only US aircraft carriers to be capable, in unmodified form, of operating the post-war generation of heavy nuclear-armed attack aircraft. All three ships eventually underwent modernisation programmes which, because they occurred over a long period of time, differed considerably in detail. After the *Franklin D. Roosevelt* was deleted in September 1977, the *Midway* was attached to the Pacific Fleet and homeported in Yokosuka in Japan, and the *Coral Sea* served as a front-line carrier with the Atlantic Fleet.

Both ships were fitted with three deck-edge aircraft lifts, but while the *Midway* had only two steam catapults the *Coral Sea* had three. A total of 1,210 tons of aviation ordnance and 0.99 million Imp gal (4.49 million litres) of JP5 aircraft fuel were carried for the air wing. The *Midway* was the more capable ship following an extensive refit in 1966, but both ships were retired early in the 1990s.

# MOSKVA CLASS

# 'Moskva' class

**Country of origin:** USSR, now CIS

**Type:** Conventionally powered anti-submarine helicopter-carrier

**Displacement:** 14,900 tons standard and 19,300 tons full load

**Dimensions:** Length 626ft 8in (191.0m); beam 75ft 6in (23.0m) and width 111ft 6in (34.0m); draught 28ft 6in (8.7m); flightdeck length 265ft 9in (81.0m) and width 111ft 6in (34.0m)

**Gun armament:** Four 57mm L/80 AA in two twin mountings

**Missile armament:** Two twin launchers for 48 SA-N-3 'Goblet' surface-to-air missiles

**Torpedo armament:** None

**Anti-submarine armament:** One SUW-N-1 twin launcher for 18 FRAS-1 missiles, two RBU-6000 12-barrel rocket-launchers, and helicopter-launched weapons (see below)

**Aircraft:** Typically 14 Kamov Ka-25 'Hormone-A' or Ka-27 'Helix-A' helicopters

**Electronics:** One 'Top Sail' 3D radar, one 'Head Net-C' air-search radar, two 'Don-2' surface-search and navigation radars, two 'Head Light-A' surface-to-air missile-control radars, two 'Muff Cob' AA gun-control radars, one 'Moose Jaw' low-frequency search and attack hull sonar, one 'Mare Tail' medium-frequency active variable-depth sonar, a very extensive ESM system including eight 'Side Globe', two 'Bell Clout', two 'Bell Slam' and four 'Bell Tap' antennae/-housings, two twin chaff launchers, and one 'High Pole-A' and one 'High Pole-B' or 'Salt Pot' (*Leningrad*) IFF systems

**Propulsion:** Four boilers supplying steam to two sets of geared turbines delivering 100,590hp (75,010kW) to two shafts

# Modern Aircraft Carriers

**Performance:** Maximum speed 31kt; range 10,360 miles (16,675km) at 18kt or 3,230 miles (5,200km) at 30kt

**Complement:** 840 excluding air group

| USSR, now CIS | | |
|---|---|---|
| Name | Builder | Commissioned |
| Moskva | Nikolayev South | May 1967 |
| Leningrad | Nikolayev South | 1968 |

**Note**
Classified by their Soviet originators with the designation *Protivolodochnyy Kreyser* (anti-submarine cruiser), the two 'Moskva' class ships are in fact hybrid helicopter carriers/missile cruisers, and were developed to counter the Western strategic missile submarines in the regional seas adjacent to the USSR. By the time that the first two vessels, the *Moskva* and the *Leningrad*, had been completed at the Nikolayev South Shipyard in 1961 and 1968, however, it was discovered that they were incapable of coping with both the number of submarines and their capabilities, so the programme was terminated.

## Moskva Class

*Lead vessel of a two-ship class, the Moskva is a hybrid carrier with potent anti-submarine and anti-aircraft armament forward and a platform aft, served by two inset lifts, for up to 14 helicopters.*

The 'Moskva' class ships were deployed primarily to the Mediterranean as part of the Soviet 5th Eskadra. They have also appeared in the North Atlantic, North Sea, Baltic Sea and Indian Ocean as part of deployed task forces or on transit.

In appearance the two ships are missile cruisers forward with extensive anti-air warfare and ASW systems located on the forward superstructure arrangement, which ends abruptly in a large steam turbine exhaust stack and main radar mast assembly. A hangar 49ft 3in (15.0m) long and suitable for two helicopters side by side is located within this structure between the stack uptakes. The after end of the ship is taken up by the flightdeck, which has four mesh-covered helicopter take-off and landing spots marked out with the numbers 1 to 4. A fifth spot, marked with the letter P, is located centrally. Two aircraft lifts, each measuring 54ft 2in (16.5m) by 14ft 9in (4.5m), serve the flightdeck from the hangar deck, which measures 213ft 3in (65.0m) by 78ft 9in (24.0m). Located under the flightdeck, this hangar can accommodate a maximum of 18 Kamov Ka-25 'Hormone-A' ASW helicopters, although 14 was the number usually carried, and these machines have been replaced by Kamov Ka-27 'Helix' helicopters.

The quintuple 21in (533mm) ASW torpedo tube mountings originally carried behind the accommodation ladders in the ship's sides have been removed. The ASW armament relies on two 6,500yd/6,000m-range 9.84in/-250mm-calibre automatically reloaded rocket-launchers and a twin SU-W-N-1 unguided ballistic missile-launcher firing the 18.6 mile/30km-range FRAS-1 rocket fitted with a 15-kiloton nuclear depth bomb as warhead. The target data for the FRAS-1 are provided by the ship's own low-frequency bow sonar and medium-frequency variable-depth sonar suites.

The 'Moskva' class ships are fitted to serve as command ships for ASW hunter-killer groups with maritime patrol ASW aircraft in support. The ships have also generated much useful data about the operation of substantial helicopter forces on major warships.

# Modern Aircraft Carriers

## 'Nimitz' class

**Country of origin:** USA

**Type:** Nuclear-powered multi-role aircraft carrier

**Displacement:** 81,600 tons standard and 91,485 tons or 96,385 tons (CVN71) or 102,000 tons (CVN72-75) full load

**Dimensions:** Length 1,092ft 0in (332.9m); beam 134ft 0in (40.8m); draught 37ft 0in (11.2m) or 38ft 9in/11.8m (CVN71) or 39ft 0in/11.9m (CVN72-75); flightdeck length 1,092ft 0in (332.9m) and width 252ft 0in (76.8m) or 257ft 0in/78.4m (CVN71-75)

**Gun armament:** Three or four (CVN70-75) 20mm Phalanx Mk 15 CIWS mountings

**Missile armament:** Three Mk 29 octuple launchers for RIM-7 Sea Sparrow surface-to-air missiles

**Torpedo armament:** None

**Anti-submarine armament:** Aircraft and helicopters (see below)

**Aircraft:** Typically 91 in a multi-role carrier air wing with 20 Grumman F-14 Tomcat, 20 McDonnell Douglas F/A-18 Hornet, 6 Grumman EA-6B Prowler, 4 Grumman KA-6D Intruder, 20 Grumman A-6E Intruder or Vought A-7E Corsair II, 5 Grumman E-2C Hawkeye and 10 Lockheed S-3A Viking fixed-wing aircraft, and 6 Sikorsky SH-3G/H Sea King helicopters

**Armour:** Belt 2.5in (63.5mm) over certain areas, plus box protection for the magazines and machinery spaces

**Electronics:** One SPS-48E 3D radar, one SPS-49(V)5 long-range air-search radar, one SPS-67(V) surface-search radar, one SPN-64(V)9 or LN-66 (CVN70/75) navigation radar, five SPN-series aircraft landing radars, one Mk 23 and six Mk 57 radars used in conjunction with three Mk 91 SAM fire-control systems, one NTDS, Links 4A, 11 and 14 data-

*Seen here in the form of the lead ship, the 'Nimitz' class aircraft carriers are the world's most-capable platforms for the projection of national power, their combination of nuclear propulsion and a very large air group (the latter supplied with fuel and ordnance from accompanying ships) offering sustained multi-role operational capability.*

links, one SLQ-32(V)4 in succession to SLQ-29(V)3 ESM system with WLR-8 warning and SLQ-17AV jamming elements, four Mk 36 Super RBOC chaff/flare launchers, one OE-82 satellite communication system, one WSC-3 satellite communication transceiver, one SSR-1 satellite communication receiver, and URN-25 TACAN

**Propulsion:** Two Westinghouse A4W or General Electric A1G pressurised water-cooled reactors supplying steam to four sets of geared turbines delivering 260,000hp (193,885kW) to four shafts

**Performance:** Maximum speed 30+kt

**Complement:** 155+2,981 plus an air group of 366+2,434 and provision for 25+45 flag staff

# Modern Aircraft Carriers

### USA

| Name | No. | Builder | Commissioned |
|---|---|---|---|
| Nimitz | CVN68 | Newport News | May 1975 |
| Dwight D. Eisenhower | CVN69 | Newport News | Oct 1977 |
| Carl Vinson | CVN70 | Newport News | Feb 1982 |
| Theodore Roosevelt | CVN71 | Newport News | Oct 1986 |
| Abraham Lincoln | CVN72 | Newport News | Nov 1989 |
| George Washington | CVN73 | Newport News | Jul 1992 |
| John C. Stennis | CVN74 | Newport News | Dec 1995 |
| United States | CVN75 | Newport News | Dec 1997 |

**Note**

The first three of the 'Nimitz' class aircraft carriers were originally designed as replacements for the elderly 'Midway' class carriers. They differ from the earlier nuclear-powered *Enterprise* in having a new two-reactor powerplant design in

Straight section of flightdeck served by two C13 Mod 1 steam catapults

Two deck-edge lifts on starboard side of flightdeck forward of the island

Hangar deck with height of 25ft 7in (7.80m)

Propulsion spaces with two A4W/A1G pressurised water-cooled reactors and heat-associated exchangers driving four sets of geared steam turbines

## NIMITZ CLASS

two separate compartments, with ordnance magazines between and forward of them. This increases the internal space available to allow the embarkation of some 2,510 tons of aviation weapons and 2.33 million Imp gal (10.6 million litres) of aircraft fuel, which are totals deemed sufficient for 16 days of continuous flight operations before stocks have to be replenished. The class is fitted with the same torpedo protection equipment as the *John F. Kennedy*, and is laid out with the same general arrangement and electronic fit.

Under the present multi-mission designations of the American carrier force, the class has been fitted with ASCAC facilities for data sharing of subsurface operations between the carrier, her escorts, their airborne ASW aircraft and supporting long-range maritime ASW and patrol aircraft. The ships also have NTDS with Links 4A, 11 and 14 intership and aircraft data-links, and are fitted with the OE-82 satellite communication system. A Tactical Flag Command Center has also been fitted for use by the fleet command officers embarked.

Four deck-edge aircraft elevators are available, two

Island superstructure for command and control, surmounted by electronic antennae including those of the SPS-10 surface-search radar, SPS-48 air surveillance radar, SPN-42 and SPN-44 carrier-controlled approach radars, and SPN-43 air traffic control radar

Tower mast carrying the antenna for the SPS-43 air-search radar

One deck-edge lift on starboard side of flightdeck abaft the island

One of four propellers

One of two rudders

# Modern Aircraft Carriers

forward and one aft of the island on the starboard side and one aft on the port side. The hangar is 25ft 2in (7.8m) high, and like those of other US carriers, it can accommodate at most only 40–50 per cent of the aircraft embarked at any one time, the remainder being spotted on the flightdeck. The angled flightdeck is fitted with three arrester wires and an arrester net for recovering aircraft. Four steam catapults are carried, two on the bow launch position and two on the angled flightdeck. A typical 'Nimitz' class air group initially comprised two squadrons of Grumman F-14 fleet-defence interceptors, two squadrons of Vought A-7 Corsair II medium attack aircraft and one squadron of Grumman A-6E Intruder all-weather medium attack aircraft, plus Grumman KA-6D Intruder tanker, Grumman E-2C Hawkeye airborne early warning, Grumman EA-6B Prowler electronic countermeasures, Lockheed S-3 Viking ASW and Sikorsky SH-3H Sea King ASW units; there were also facilities for a Grumman C-2A Greyhound carrier on-board delivery aircraft. The current aircraft fit is detailed above.

In 1981 the first of at least five improved 'Nimitz' class carriers was ordered after much discussion both within the Congress and the Pentagon. These vessels have Kevlar armour (retrofitted in the earlier ships) over their vital areas, and have additional hull protection arrangements. Under normal usage, the core life of the A4W reactors is expected to provide a cruising endurance of about 13 years before replacement. Although the class is relatively new, it is

# Nimitz Class

*The Dwight D. Eisenhower (above and below) is the second unit of the exceptionally powerful 'Nimitz' class, whose multi-role capability is based on an air group of some 90 aircraft operating from a very large flightdeck with straight and angled sections each served by two C13 Mod 1 steam catapults. The flightdeck has four deck-edge lifts (one to port and three to starboard), and the angled section of the flightdeck has four sets of Mk 7 Mod 3 arrester wires.*

planned for the 'Nimitz' class to undergo SLEP refits during the first decade of the next century in order to extend their service life into the 2020s.

These ships are undoubtedly the world's most powerful surface combatants, and a total of eight or nine is planned to provide the backbone of the US Navy's carrier battle groups until well into the next century.

# Modern Aircraft Carriers

## 'Principe de Asturias' class

**Country of origin:** Spain

**Type:** Conventionally powered light aircraft carrier

**Displacement:** 14,700 tons standard and 16,700 tons full load

**Dimensions:** Length 642ft 9in (195.9m); beam 79ft 9in (24.3m) and width 105ft 0in (32.0m); draught 30ft 10in (9.4m); flightdeck length 574ft 0in (175.0m) and width 88ft 7in (27.0m)

**Gun armament:** Four 20mm Meroka CIWS mountings

**Missile armament:** None

**Torpedo armament:** None

**Anti-submarine armament:** None

**Aircraft:** Typically 24, with between 6 and 12 BAe/McDonnell Douglas AV-8B Harrier II fixed-wing aircraft, plus between 2 and 4 Agusta (Bell) AB.212, between 6 and 10 Sikorsky SH-3 Sea King, and 2 Sikorsky S-70L Seahawk helicopters

**Electronics:** One SPS-52C/D 3D air-search radar, one SPS-55 surface-search radar, one SPN-35A aircraft landing radar, one RTN 11L/X missile-warning radar, four VPS-2 Meroka fire-control radars, one NTDS, Links 11 and 14 data-links, one Nettuno ESM system with warning and jamming elements, one SLQ-25 Nixie towed torpedo-decoy system, two Prairie/Masker hull/propeller blade noise suppressors, four Mk 36 Super RBOC chaff/flare launchers, and URN-22 TACAN

**Propulsion:** COGAG (COmbined Gas turbine And Gas turbine) arrangement with two General Electric LM2500 gas turbines delivering 46,400hp (34,600kW) to two shafts

## PRINCIPE DE ASTURIAS CLASS

**Performance:** Maximum speed 26kt; range 7,455 miles (12,000km) at 20kt

**Complement:** 90+465 plus 208 air group and flag staff

| Spain | | | |
| --- | --- | --- | --- |
| Name | No. | Builder | Commissioned |
| Principe de Asturias | R11 | Bazan | May 1988 |

**Note**
To replace the obsolete *Dédalo*, in June 1977 the Spanish ordered a gas turbine-powered vessel based on the final design variant of the US Navy's abortive Sea Control Ship. Named *Principe de Asturias*, the new ship has a flightdeck fitted with a 12° 'ski-jump' ramp blended into the bow. Two aircraft lifts are fitted, one of them at the extreme stern. For the air wing of this vessel, Spain ordered the McDonnell Douglas AV-8B Harrier II STOVL fighter and the Sikorsky SH-60B Seahawk ASW helicopter. The aircraft complement can be increased to a total of 37 machines in time of war.

*The* Principe de Asturias *is a good example of the type of small platform now utilised by several navies for the operation of helicopters and STOVL warplanes launched at a high weight with the aid of a 'ski-jump' ramp at the front of the flightdeck.*